Modular Science

MAN & OTHER ANI

Valerie Wood-Robinson

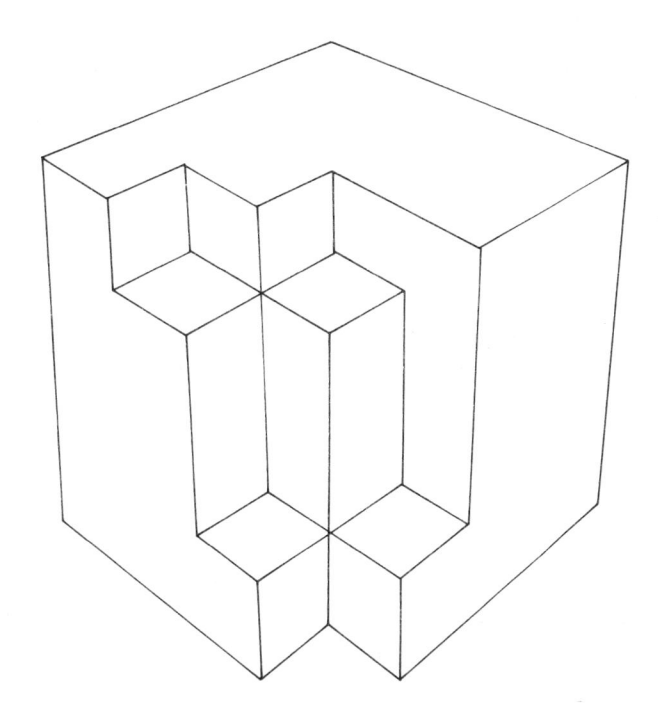

Blackie

Glasgow and London

Acknowledgments

The author and publishers wish to thank the following for permission to reproduce copyright material:

M. C. Wilkes (page 4, left/page 24, middle), Tom Leach (page 5, far left), N. Jagannathan (page 5, middle right), E. K. Thompson (page 5, far right/page 21, left), A. J. Bond (page 20, left), W. S. Paton (page 24, left), page 29 — all supplied by Aquila Photographics; Leslie Jackman (page 4, right/page 6); K. V. Crocket (page 5, middle left/page 16/page 19, left/page 19, right/page 21, middle/page 21, right); Keystone Press Agency Ltd (page 7/page 32, middle left/page 32, top right/page 32, bottom right); L. Hugh Newman, Natural History Photographic Agency (page 12/page 17, left/page 17, right); Syndication International Ltd (page 14/page 32, top left/page 32, bottom left); John Topham Picture Library (page 20, middle); Ian Beames and Ardea, London (page 20, right); British Aerospace, Filton (page 24, right).

Modular Science

This series was planned and developed with the assistance of an Editorial Panel. Its members are

Bob Fairbrother, Lecturer in Science Education, Chelsea College
Edgar Jenkins, Senior Lecturer in Science Education, University of Leeds
Peter Scott, Headmaster, City of Leeds School

ISBN 0 216 90588 5

First published 1981

Copyright © Valerie Wood-Robinson 1981

All rights reserved. No part of this publication may be reproduced, stored in a retrieval system or transmitted in any form or by any means, electronic, mechanical, photocopying or otherwise without prior permission from the publisher.

Published by Blackie and Son Ltd
Bishopbriggs, Glasgow G64 2NZ
Furnival House, 14–18 High Holborn, London WC1V 6BX

Filmset by Filmtype Services Limited, Scarborough
Printed in Great Britain by Bell & Bain Ltd, Glasgow

CONTENTS

*The work in the numbered chapters of this book is called the core.
Your teacher will probably want you to work all the way through it.*

CHAPTER 1
The animal kingdom

CHAPTER 2
Moving and responding

CHAPTER 3
Getting energy

CHAPTER 4
Growing and reproducing

When you have studied these chapters you will know how to use the word 'animal' as a scientist would. You will have an idea of the variety of animal life. You will know why scientists place animals in groups and you will understand some of the problems of doing this. You will know that you perform the basic living activities like other animals. You will also realize that each kind of animal has its own special way of doing these activities.

Your teacher will then want you to do one or more of the following options:

OPTION A
Ways of feeding

OPTION B
Animal Olympics

OPTION C
Investigating your breathing

OPTION D
Measuring growth

When you have learnt more about the variety of animal activity and more about yourself, you will be ready to understand some more ideas about classifying animals.

CHAPTER 5
Man's place in the animal kingdom

After reading this you will be able to discuss the similarities and differences between Man and Other Animals

CHAPTER 1 — The animal kingdom

Did you know that there are more than a million different *kinds* of animal alive in the world?

Horse and foal

Pied wagtail

Each distinct kind of animal is called a SPECIES. Man is one species of animal. Every person is an individual, but they are all humans. When a man and a woman produce a baby, it is always a human baby, not a puppy or a monkey. The word 'man' is used to mean the species of human beings, as well as meaning one adult male person.

Another species of animal is the dog. There are many different varieties of dog, but they can all breed together and they produce puppies, not kittens or parrots. A dog cannot breed with a cat, nor with any other species. An animal can breed only with its own species and produce young ones of the same species. Members from different species cannot breed together successfully.

Scientists from different countries, e.g. English, Russian, Italian and Japanese scientists, might not know the word for 'dog' in each other's language. So each species has a scientific name which is known all over the world. The scientific name for the dog species is *Canis familiaris*; for man, *Homo sapiens*; and for the housefly, *Musca domestica*.

ACTIVITY 1.1
Make a table like the one below. Write a list of 10 animals in the first column. If there is a special word for the male, female or baby of the species, fill that in. Find out the scientific name for each species if you can.

Species of animal	Male	Female	Baby	Scientific name
Man	man	woman	baby	*Homo sapiens*
African elephant	bull	cow	calf	*Loxodonta africana*

What is an animal?

Look at the list of animals which you made for Activity 1.1. How do you know that they are all animals? Have you put any insects, e.g. butterflies, on your list? About half the species of living animals (nearly ¾ million species) are insects. Have you put any birds on your list? What about fish? What about worms and snails? All these creatures are animals. We tend to think that only MAMMALS are animals. Mammals are animals which are most like ourselves.

ACTIVITY 1.2
Think of as many different animals as you can, including some which are very different from you and from each other. The pictures in this book will give you a start. You can get some more ideas from library books, or from films or an exhibition which your teacher may show you.

Record your ideas by writing a list of different animals, or by drawings, or collect pictures from old magazines. Keep your record, which we shall call your 'collection', to use in some other Activities in this book.

Chimpanzee   Common toad

Why are all the things in your list or collection called animals? What does the word 'animal' mean? It means these things:
1 An animal is a living thing. (Of course living things die, so some of your examples may be the remains of dead animals.)
2 An animal feeds by taking food into its body from its surroundings.
3 Most animals can move about from one place to another place. Some animals stay fixed in one place, but they can move parts of their bodies by themselves.

Animals form one of the KINGDOMS of living things. The other is the PLANT KINGDOM. Plants are usually green. They make their food. They cannot move, except by growing or by being pushed. All living things are made of tiny 'building bricks' called CELLS. But the cells of plants are different from the cells of animals.

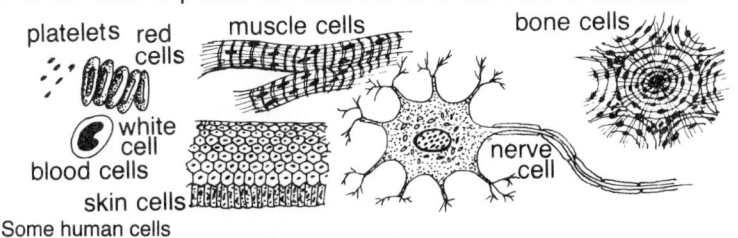

Some human cells

Classifying animals

As you can see, there are lots of different animals. Scientists divide the ANIMAL KINGDOM into groups of similar species. You already know the names of some groups: insects, birds, mammals.

CLASSIFICATION means arranging species into groups, according to their similarities. It is easier to study animals if they are classified. The group name is a kind of shorthand, full of information. It tells you a lot about a species even if you do not know that particular animal. If you read that a lumpsucker is a fish, you immediately know quite a lot about a lumpsucker.

Cobra   Swallowtail butterfly

ACTIVITY 1.3

Try to classify the animals in your collection. Arrange all the animals into sets of similar ones. Some sets may have only one member if you cannot find another animal enough like it for that set. Give each set a name which describes the group. You need not know all the scientific names for the groups at this stage. Compare your classification with that of other people.

When you classify animals, your sets may not match those of other people. This does not mean that anyone is 'wrong'. There are many different ways of classifying animals. Each person may have thought of a different feature to be important for his classification.

CHAPTER 2
Moving and responding

Animal movement

Animals move by their own power. Most animals can move their whole body from place to place, but some animals do not move around at all. They are called SEDENTARY animals. They can move parts of their bodies. A sea anemone is a sedentary animal which moves its tentacles to help it catch food.

Sea anemone

ACTIVITY 2.1
Put an earthworm into a glass tube. Watch how it moves. Notice which parts get fatter, thinner, shorter, longer, at each stage of its movement. Make a series of drawings to show how the worm changes shape as it moves.

Put the earthworm on a piece of rough paper. Put your ear near to the paper and listen to the worm moving.
Pick up your worm. Gently rub your finger along its underneath side, from front to back. Then do it from back to front. Look at this underneath surface through a hand lens. Write about:
1. what you could hear;
2. what you could feel;
3. what you could see;
4. what the earthworm has on its underside;
5. how these structures help the worm to move.

Try putting the worm on different surfaces, e.g. polythene, glass, tile. Can it get a grip on all the surfaces? Which surfaces can it move along most easily?
Does the earthworm have legs? Do you think it has a backbone?

An earthworm is an INVERTEBRATE, which means an animal without a backbone. It has no bones in its body to keep it firm. Its body is a tube full of liquid. The liquid keeps it firm, rather like a plastic bag full of water. The wall of the tubular body is made of MUSCLES. Muscles have the special job of CONTRACTING (getting shorter). When muscles contract they pull on other parts of the body, making these parts move. The opposite of contracting is RELAXING, when the muscle goes back to its starting length. Alternate contraction and relaxation of a worm's muscles make it move along. Look at this diagram to help you understand:

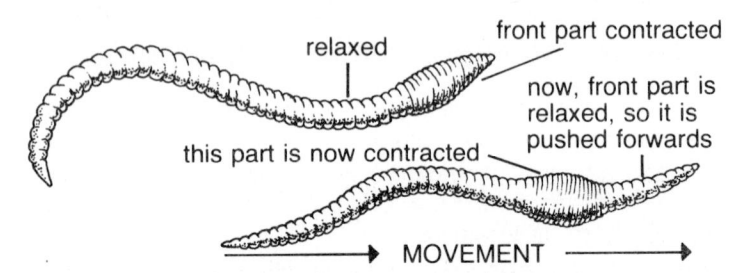

relaxed — front part contracted
this part is now contracted — now, front part is relaxed, so it is pushed forwards
→ MOVEMENT →

All animals have muscles for movement. The muscles are usually arranged in two sets which work in opposite ways. While one set contracts, the other set relaxes, and vice-versa. They are called ANTAGONISTIC muscles. When a muscle contracts it gets shorter, bulges out and feels harder to touch.

Contracting shoulder muscles

ACTIVITY 2.2
Do each of these actions in turn: bend your knee, straighten your leg, stand on tip-toe to raise your heel. Feel each muscle with your hand. For each action: Which muscle is contracting? Which is its antagonistic muscle? What does the contracting muscle pull on?

There are many muscles in your legs, your back, your bottom and even your arms, which help you to walk. Your muscles pull on your rigid bones. This makes your skeleton bend at JOINTS like your knees and hips. Your legs push against the floor and this makes you move forwards.

You are a VERTEBRATE, an animal with a backbone. Your bony skeleton is important for your kind of movement. Even invertebrates have some sort of skeleton to help them move. The liquid in an earthworm's body does this job. Insects are invertebrates, but they have a firm skeleton on the outside of their bodies to help them move.

Responding to stimuli

Animals move because they are influenced by something in their surroundings. A lion may chase a zebra for food, a slug may crawl under a dark stone, a fly will fly faster on a hot day. Anything which makes you (or an animal) move is called a STIMULUS (plural STIMULI). Your movement is called a RESPONSE. Most responses involve movement of all or part of an animal's body.

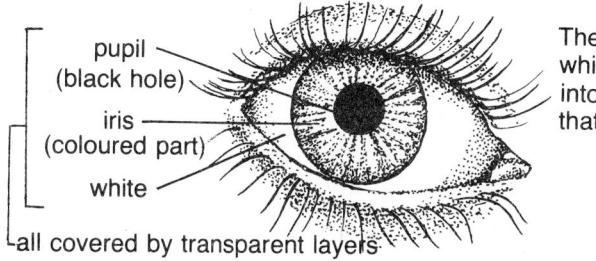

pupil (black hole)
iris (coloured part)
white
all covered by transparent layers

The pupil is a hole which lets light into your eye so that you can see

ACTIVITY 2.3
Work with a partner in a dim room. Look at the pupils of your partner's eyes. Then shine a low voltage light on his eye for a few seconds. Do *not* use a bright light and do *not* shine it for more than a few seconds. Watch what happens to his pupils.
You can do this experiment on yourself, looking in a mirror. What response did the pupil make? Can a hole really make a response? Which part of the eye moved in this response? Does this part contain muscles? Can you deliberately control this response, or does it happen automatically?

ACTIVITY 2.4
Again you need a darkened room. You can study the responses of a maggot. It is the young stage, or larva, of a blowfly. It is an insect.

Arrange a narrow beam of light to shine across a large sheet of white paper. Put a maggot in the beam. Watch what the maggot does. Put a drop of red ink on its 'tail' so that it marks a trail on the paper. Try putting the maggot sideways to the light. Then try it 'head' towards the light. Investigate the animal's response to two beams arranged at right angles.

Investigate the response of another maggot trailing a different colour of ink. You could run a 'maggot Derby'!

Write down what you discover about the responses of maggots to the stimulus of light.

You may try variations on this experiment (without harming the maggots) to test any ideas you may have about other stimuli, such as heat from the lamp, wetness from the ink trails, etc.

Maggots make quite simple responses to a few stimuli. You respond in more complicated ways to many different and detailed stimuli. You are responding *now* to the complicated stimulus of marks on paper which we call printed words. Maggots have simple SENSE ORGANS to detect stimuli. You have complex sense organs, including eyes, ears, taste buds and skin. Some of your responses are automatic, unconscious REFLEXES. You saw the pupil reflex in Activity 2.3. Other responses are conscious and deliberate. A part of your body which performs a response is called an EFFECTOR ORGAN.

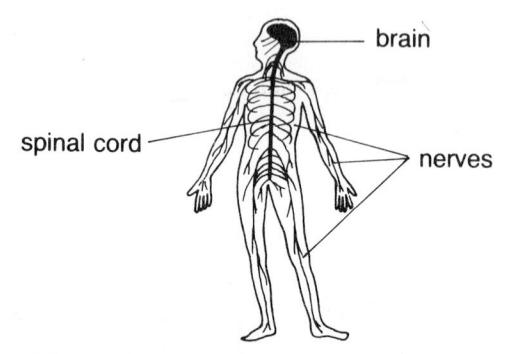

Your nervous system

ACTIVITY 2.5
Make a table, like the one below, about responses which *you* make to stimuli.

Stimulus	Sense organ	Response	Effector organ	Is it a reflex?
1. smell of food	nose	mouth waters	salivary glands in mouth	reflex
		walk to dining room	muscles of legs	not reflex

Fill in some more examples.

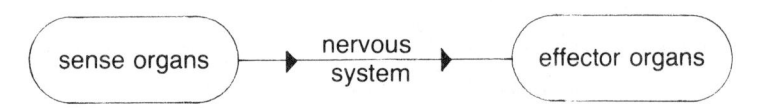

sense organs → nervous system → effector organs

All animals have a NERVOUS SYSTEM to carry messages from their sense organs to their effector organs. This link enables them to give the correct response to a stimulus. The nervous system of a maggot or a worm is simple compared with yours. Man has the most developed brain of all the animals. This enables him to think and to plan and perform complicated responses.

CO-ORDINATION means linking all the parts of the body to work together. When you, or any other animals, walk, the nervous system makes sure that all the muscles work together and in the correct order. Your nervous system is one of the systems which co-ordinates you. The other way in which your activities are co-ordinated is by hormones, or chemical messengers. These control on-going activities like growth and development, which are the subject of Chapter 4.

CHAPTER 3

Getting energy

All movement needs a supply of energy. You get all your energy supplies from the food you eat. All animals have to eat living or dead food to obtain energy supplies. Green plants can 'capture' the energy of sunlight and store it in the food they make. The food made by plants feeds all animals, either directly or indirectly.

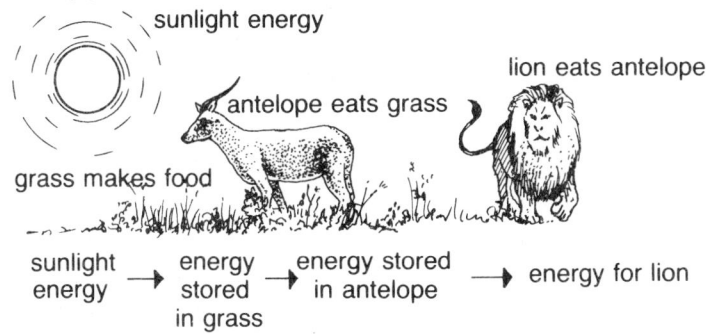

sunlight energy

antelope eats grass

lion eats antelope

grass makes food

sunlight energy → energy stored in grass → energy stored in antelope → energy for lion

An energy chain

The way that energy is passed on in food can be shown by a FOOD CHAIN or ENERGY CHAIN. Every chain must start with a green plant capturing sunlight energy. In the sea and ponds, microscopic green plants are at the start of most chains.

ACTIVITY 3.1
Write energy chains, starting with a plant capturing sunlight energy, to show how each of these animals get their energy for movement: a horse; a cat; an Eskimo man who lives mainly on fish; a Masai tribesman who lives mainly on the products of his cattle.

Food contains energy

The food we eat consists of many different chemical substances, including PROTEINS, FATS and CARBOHYDRATES. We get most of our energy from foods rich in carbohydrate or rich in fat. If you eat more of these 'energy foods' than you need, your body changes the surplus into human fat. This is stored, making you look fat. Exercising is a healthy way to slim. The fat stores of the body are broken down to provide energy for the exercise.

ACTIVITY 3.2
You can show that a peanut contains energy, and measure how much it contains. Follow the diagrams. Does the water get hotter? What is the rise in temperature? Where does the energy come from to heat the water?
Work out the number of joules of energy which were stored in the peanut, as follows.

$$\text{joules of energy in one peanut} = \text{rise in temperature (in °C)} \times 20 \times 4.2$$

(20 = mass of 20 cm³ of water)
Is this an accurate result? How could you improve the experiment to get a more accurate result?

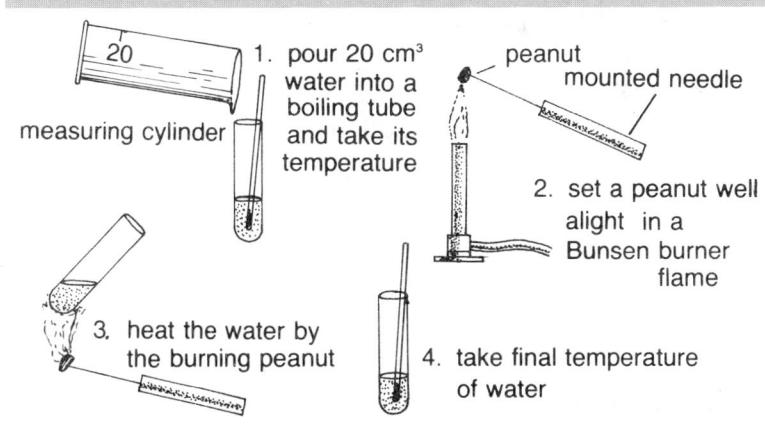

1. pour 20 cm³ water into a boiling tube and take its temperature

measuring cylinder

peanut mounted needle

2. set a peanut well alight in a Bunsen burner flame

3. heat the water by the burning peanut

4. take final temperature of water

eyelid muscles need energy for blinking

nervous system needs energy to carry messages

heart muscles need energy to pump blood

heat energy released from food keeps the body warm

kidneys need energy to do their work of controlling body liquid

leg muscles need energy for walking

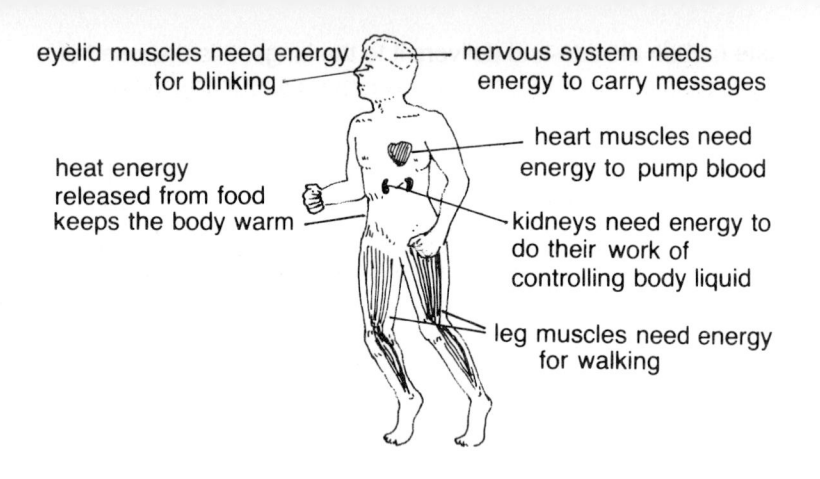

Energy is released from food in all the cells of your body, and in the cells of every living thing.

How energy is released from food

The peanut in Activity 3.2 burnt by combining with oxygen in the air. This burning or OXIDATION, gave out heat and light energy. It produced a waste gas called CARBON DIOXIDE. In all living cells, a similar oxidation process is going on all the time. It is called RESPIRATION. It does not need such a high temperature as burning. Food, in the form of GLUCOSE, combines with oxygen. Energy, as heat, or movement, etc. is released from the glucose.

The food you eat consists of many different chemicals. The cells of your leg muscles, brain and all parts of your body need energy food in the form of GLUCOSE. Food is digested to glucose in the intestines.

ACTIVITY 3.3

Complete a table (like the one at the top of the next column) which shows how respiration is similar to combustion (burning).

		Respiration	Combustion	
1	Where does it happen	in all living cells	in a fireplace	in a car engine
2	What is the fuel?	glucose (food)		
3	What type of process?	oxidation (combining with oxygen)		
4	What is the main waste gas?			
5	What form of energy is released?			

Transport in your body

Every cell of your body needs deliveries, messages and waste disposal, just as your household does. In the body there is one 'delivery service', the BLOOD. Your HEART is a pump which circulates your blood through a continuous system of tubes called blood vessels. As it goes round your body, your blood picks up things where they are plentiful and drops them off where they are needed.

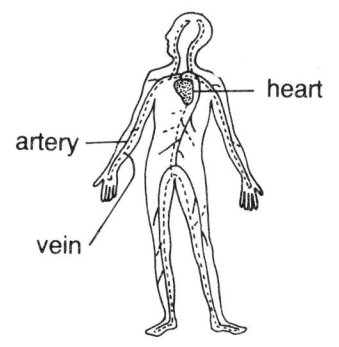

Your circulatory system (much simplified)

Glucose from digested food is picked up at your small intestine and delivered to all the respiring cells. Food provides building materials for growing and repairing your body, as well as providing energy. Soluble body-building food is carried in the blood to cells which need it.

ACTIVITY 3.4
Watch your teacher dissecting a small mammal. Look for the parts which have been mentioned in this book. Answer the questions which your teacher may set you.

All vertebrates have a circulatory system with a heart pumping blood round it. Earthworms and octopuses are some invertebrates which have circulatory systems but with simpler hearts. Insects have blood but it is not enclosed in tubes. It fills the insect's body and surrounds all the cells.

Oxygen and carbon dioxide transport
Your blood picks up oxygen at the lungs and delivers it to all the respiring cells. As the blood goes past these cells it collects the waste carbon dioxide and delivers it to the lungs. It is disposed of in your exhaled air. (Option C tells you more about breathing.)

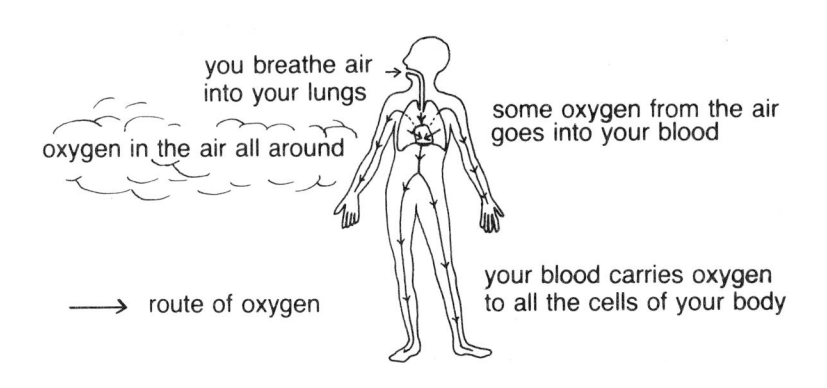

ACTIVITY 3.5
Make 3 outline drawings of a person like the one shown here. On the first, mark the paths of oxygen copied from the diagram. On the second, mark the path of carbon dioxide away from the cells. On the third, mark the path of food to the cells. You cannot draw the detailed paths accurately, but show the overall routes with different coloured lines.

Getting energy in the animal kingdom

All animals obtain energy by respiration in their cells in exactly the same way as you do.

glucose + oxygen ⟶ ENERGY + carbon dioxide + water

However, different animals have different ways of getting food and oxygen to their cells and of getting rid of waste products. Option A is about the many ways of feeding in the animal kingdom. There are also many ways of exchanging gases with the outside air, or with the water surrounding the animal.

ACTIVITY 3.6
Put a goldfish or other small fish in a suitable tank of water. Watch it breathing. Look at the movements of its mouth and OPERCULUM. Work out how water is pumped through. How many times does the fish breathe in one minute? Time the breathing rate again, after adding some warm water to the tank to raise the temperature by about 5°C.

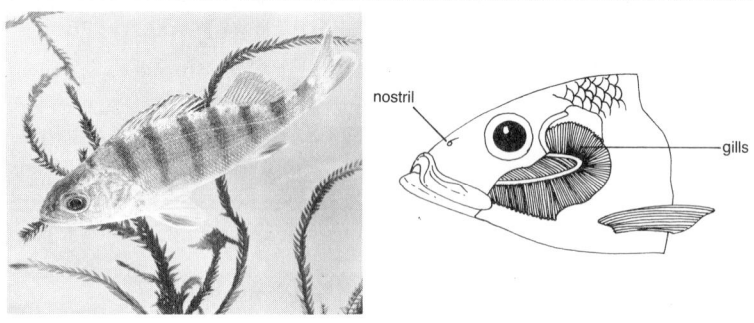

Perch

ACTIVITY 3.7
Look at the breathing organs of a dead fish. Pull on the mouth and operculum to imitate the breathing movements of a live fish. Cut off the operculum on one side to see the gills. Why do the gills look red? Notice that the skin on the gills is very thin. Why is this important? Why can a fish not breathe out of water?

Getting rid of waste products

Carbon dioxide would poison an animal if it stayed in its body. So all animals have to EXCRETE carbon dioxide. EXCRETION involves getting rid of other poisonous waste products as well. UREA is a waste product made in the liver, from surplus food protein and broken old cells. The blood delivers this urea to the kidneys, along with some water, to make URINE. The kidneys have to work hard to make the urine the correct concentration, so that you do not lose too much water. Birds, reptiles and insects excrete solid uric acid, instead of urea dissolved in water as we do. In this way, they avoid losing water. As a result they don't have to drink much.

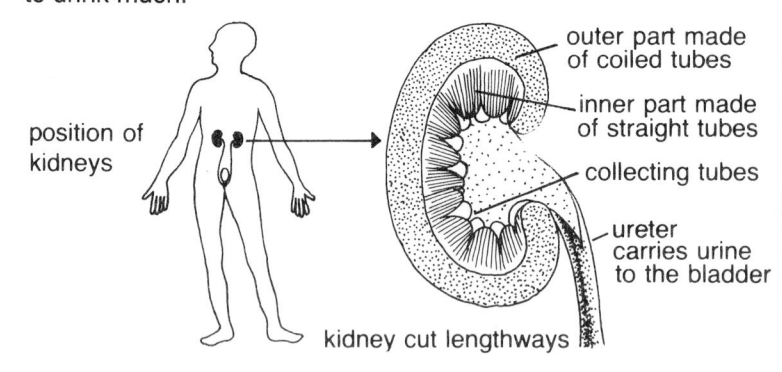

ACTIVITY 3.8
Look at a kidney. Remove the thin covering membrane. What colour is the kidney? What makes it this colour?
Cut the kidney lengthways into two equal halves. Follow the collecting tubes into the kidney. Can you see that the kidney is made of thousands of tiny tubes?

ACTIVITY 3.9
Try to find out more about the work of the kidneys. What would happen if your kidneys were faulty and you kept too much water in your body? What would happen if you had too little water in your body? Why do you need to pass urine more often in cold than in hot weather? How do the kidneys of the desert mammals, like gerbils, differ from ours?

Animals cannot digest and use all the food they eat. The undigested food is eliminated as droppings or FAECES. The droppings of birds are a mixture of undigested food, such as blackberry pips, which is eliminated, and white uric acid, which is excreted from the kidneys. Birds have only one opening for both of these functions and for reproduction.

CHAPTER 4

Growing and reproducing

You started life as one tiny cell, about the size of a pin prick. Now you are a large animal, made of millions of cells. You have grown and you are probably still growing. GROWTH means a permanent increase in size. It involves making new cells. Even if you have reached your adult height, parts of your body go on growing all your life. Some of your blood cells and skin cells are dying all the time. New cells grow to replace the dead ones. If you cut your skin or break a leg bone, new growth repairs these damaged tissues.

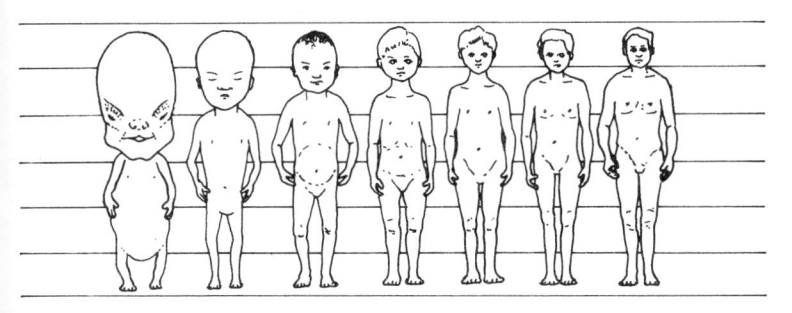

Changes in proportions from before birth to adult

ACTIVITY 4.1
Fill in the missing words about your growth.
1. My .... is growing all the time, and needs cutting.
2. The part which has grown least since I was born is my ....
3. To judge whether I am still growing, I could regularly measure my ......
4. If I damage my ....., it can easily repair itself by growth.
5. A part which could not mend itself if it was damaged, is my ......
6. I will probably reach my adult height when I am .... years old.

Growing in stages
A new-born baby is obviously a person. A newly-hatched fish is a miniature adult fish. They grow and develop steadily to their adult size. But some animals hatch as LARVAE. This means that they look very different from an adult of their species. They have to make big changes as they grow, to become an adult.

The change from the larva to the adult is called METAMORPHOSIS. All insects have a larval growing stage, with metamorphosis to the adult stage. This is partly because their skin is hard. It is the skeleton on the *outside* of their body.

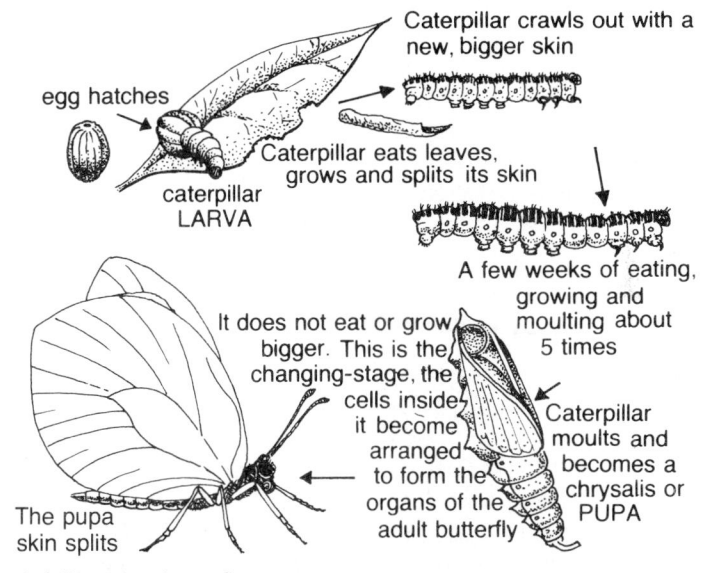

- egg hatches
- caterpillar LARVA
- Caterpillar crawls out with a new, bigger skin
- Caterpillar eats leaves, grows and splits its skin
- A few weeks of eating, growing and moulting about 5 times
- Caterpillar moults and becomes a chrysalis or PUPA
- It does not eat or grow bigger. This is the changing-stage, the cells inside it become arranged to form the organs of the adult butterfly
- The pupa skin splits
- A fully grown butterfly comes out

13

ACTIVITY 4.2
Find out the ages at which these animals reach their adult sizes: dog, cat, gerbil, rabbit, elephant, horse, tortoise, salmon, housefly, mosquito, frog, sparrow. In each case, is it the same as the age they can begin breeding?

How big will you grow?

Food provides building materials for growing and repairing your body. It also provides energy for the work of growing. Proteins are the main body-building chemicals. You get these from meat, fish, cheese, eggs, etc. Your diet also supplies other chemicals which you need for healthy growth. It supplies calcium for strong bones and teeth, and iron for blood. A person cannot grow to his maximum size if his diet is poor. But however much you eat, you cannot grow bigger than the adult size that is 'programmed' for you in your inheritance. If you come from a tall family you will probably be tall. This is because all your cells contain GENES for tallness, inherited from your ancestors.

Growth is regulated by HORMONES (chemical messengers). The PITUITARY GLAND, in the head, makes growth hormone. In some people this gland makes the wrong amount of hormone. Too much growth hormone makes a person grow to be a giant. Too little makes a person a dwarf.

Hormones control the growth of all animals. Insects have hormones which control moulting and metamorphosis. Tadpoles make a hormone called THYROXINE. This causes them to metamorphose into frogs. We also make thyroxine which controls our growth and development.

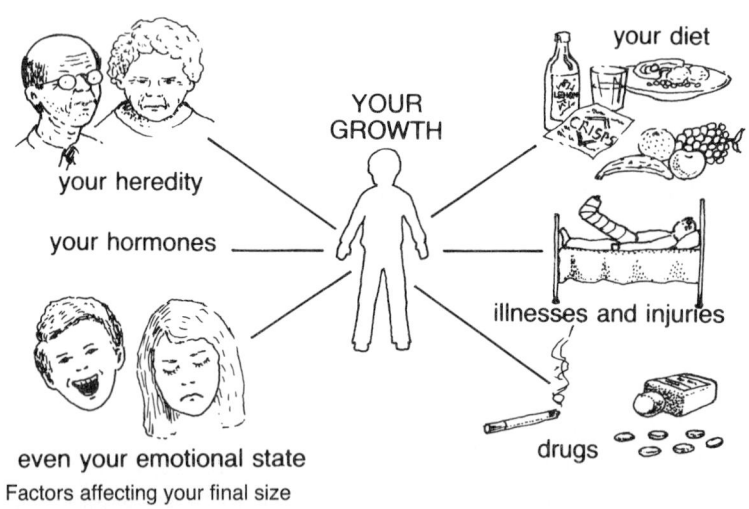

Factors affecting your final size

ACTIVITY 4.3
A person grows fastest and develops most *before* he is born. The food and drugs which an expectant mother takes, influence how her baby grows.
Make a poster to show DOs and DON'Ts about food and drugs for a pregnant woman.

Growing new animals

Some animals are made of only one cell. AMOEBA is a one-celled animal. When it grows to its maximum size, the cell divides into two separate cells. Each is a new amoeba, so one parent has split into two young ones.

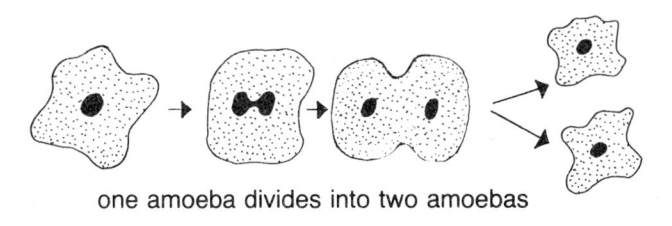

one amoeba divides into two amoebas

HYDRA is an animal which can grow a young hydra as a branch of its body. It is a small, many-celled animal which lives in ponds. It is a COELENTERATE, related to sea anemones. When it is well-fed, a bump grows on its body. This develops tentacles and a mouth like the parent. Then it breaks off the parent and becomes a separate young hydra.

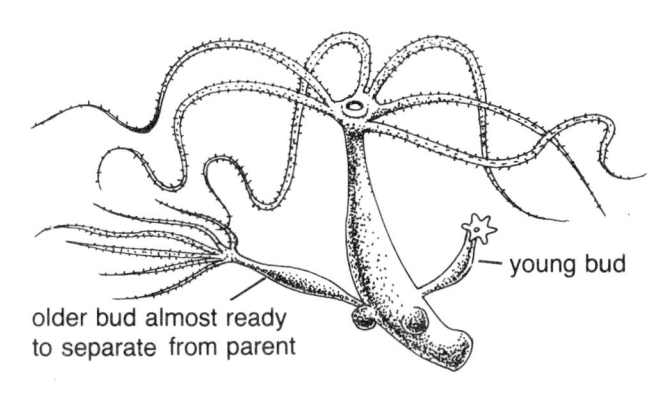

young bud
older bud almost ready to separate from parent

Many other small animals can split off parts of themselves to form young ones. Some species can reproduce in this way if they are accidentally cut into two or more pieces. When one parent can produce young ones by this sort of growth, the process is called A SEXUAL REPRODUCTION (non-sexual reproduction).

Sexual reproduction

Only simple animals can reproduce by growing branches or by splitting. Most animals produce young ones by SEXUAL REPRODUCTION. Animals cannot breed as soon as they are born. Humans reach PUBERTY in their early teens. Our bodies are capable of producing children from that age. Most people are not fully grown nor old enough to bring up a family when they reach puberty.

From the age of puberty, animals make special reproductive cells called GAMETES. The gametes made by male animals are called SPERMS. The gametes made by female animals are called EGGS or OVA. A baby animal starts when one sperm joins with one egg. The joining of a sperm with an egg is called FERTILIZATION.

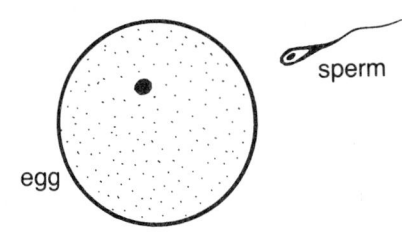

sperm
egg

Humans, dogs, mice and all other mammals, have INTERNAL FERTILIZATION. The father puts millions of his sperms into the mother, with his erect penis, during sexual intercourse. A sperm fertilizes an egg *inside* the mother's body. The fertilized egg grows into a baby inside the mother's UTERUS (WOMB). Food and oxygen are filtered from the mother's blood into the baby's blood by the PLACENTA, which fixes the baby inside the uterus. After the baby is born, the mother feeds it with milk from her MAMMARY GLANDS. The parents look after their young ones in many other ways.

hair develops on face and body at puberty

voice breaks at puberty

testes make sperms

penis puts sperms into woman

breasts (mammary glands) develop at puberty

ovary makes eggs (ova)

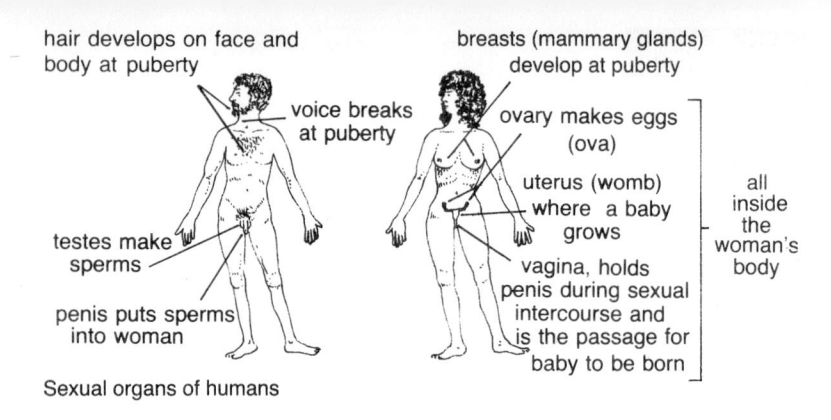

uterus (womb) where a baby grows

vagina, holds penis during sexual intercourse and is the passage for baby to be born

} all inside the woman's body

Sexual organs of humans

ACTIVITY 4.4
Choose a mammal (not human) to write about. Describe its reproduction. Here are some questions you might think about:
1. At what age does the animal first breed?
2. Do the parents live together after mating?
3. What is the gestation period (length of pregnancy)?
4. How many offspring are there likely to be in the litter?
5. How well-developed are they at birth?
6. How long is the period of parental care?

Tortoises mating

Eggs with shells

Mammals are the only animals which develop their babies in a womb, then give birth. Like mammals, birds mate and have internal fertilization; so do reptiles. But in these animals, when an egg has been fertilized inside the mother, it gets a shell around it. Then the egg is laid and the baby animal develops inside the egg outside the mother. Birds' eggs have hard shells. Reptiles' eggs have leathery shells. When the babies are developed enough they hatch out.

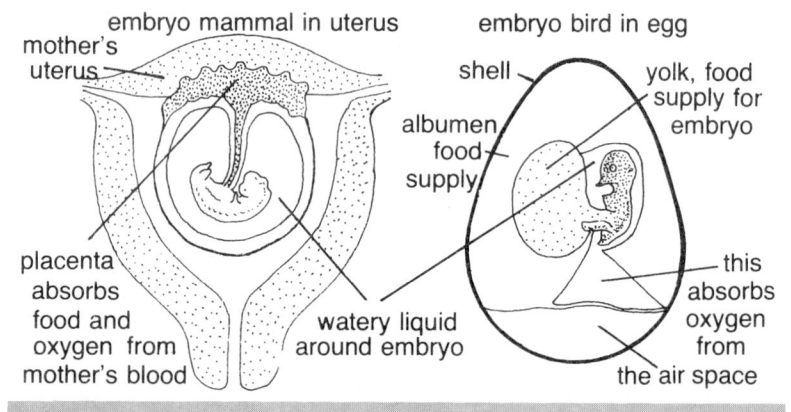

embryo mammal in uterus

mother's uterus

placenta absorbs food and oxygen from mother's blood

watery liquid around embryo

embryo bird in egg

shell

yolk, food supply for embryo

albumen food supply

this absorbs oxygen from the air space

ACTIVITY 4.5
Make a chart like this and fill in the spaces.

	Human embryo (mammal)	Sparrow embryo (bird)	Tortoise embryo (reptile)
How is it protected?			
How is it fed?			
How does it obtain oxygen?			
How does it get out?			

Reproduction in water

All animals need water for fertilization and development. In mammals, birds and reptiles, this water is inside the mother animals and inside egg shells.

Insects, spiders and scorpions have to mate. Internal fertilization then occurs in the water inside the female animal.

Animals which live in water can have fertilization outside their bodies. This is called EXTERNAL FERTILIZATION. It is more uncertain than the internal method. A male fish pours millions of sperms into the sea or river. A female fish pours large numbers of ova into the water. The parents swim away. By chance a sperm may meet an egg of the same species and fertilize it, but most of the gametes die. Fertilized fish eggs develop in the water. They do not have shells, but they have a protective jelly round them.

Sticklebacks, with male stimulating female to lay eggs in nest

Each species of animal has its own detailed way of breeding, which may be rather different from the general pattern for its group. Sticklebacks are fish, but the male courts the female by special swimming movements, then the pair build a nest for their eggs. Dogfish have internal fertilization, and the egg is laid in a case. The duck-billed platypus is a mammal which lays eggs.

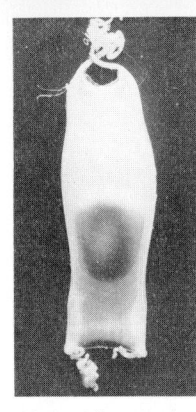

'Mermaid's purse', really a dogfish egg case

Duck-billed platypus

Most species have separate male and female animals. Some invertebrates are HERMAPHRODITE. This means that each individual has both male and female organs. An earthworm has TESTES, making sperms, *and* OVARIES, making ova, in its body. Even so, earthworms mate and exchange sperms. One cannot fertilize its own eggs.

The life cycle

When a young animal is born or hatched, it has to grow and develop, until it is adult enough to reproduce its own young. Some animal parents care for their young and help them to become independent adults. Other species leave their young ones to develop on their own. The time needed to grow up varies from a few days, needed by flies, to many years for humans.

ACTIVITY 4.6
At what age are humans 'grown-up'? Write your ideas about this, or arrange a class discussion. Think about the age at which you become adult physically, mentally, emotionally, financially, legally. Why is it more difficult to decide when a person is adult than to know whether a dog or a fly is adult?

OPTION A — Ways of feeding

All animals get their food by eating things which are alive or parts of things which have been alive. Some animals eat only plants. These animals are called HERBIVORES. They have to move to where the plants are, but they don't have to move far or fast.

Animals which eat meat from other animals are called CARNIVORES. They usually have to move fast to catch their food. The cheetah, which is the fastest animal on land, is a carnivore. Humans are OMNIVORES because we eat both plants and animals. The whole way of life of an animal is related to what it eats.

Some herbivores,   Some carnivores

ACTIVITY A.1
Choose one herbivore and one carnivore that feeds on it (e.g. zebra and lion). Make a table to compare their feeding habits and ways of life. Think about what each eats, how it gets its food, how it spends its time, its family life, its senses, its movements and its protection.

Biting
An animal's teeth and jaws are suited to its diet. All teeth break up food, but some are specialized to break it in special ways. Beavers' front teeth work like chisels, dogs' side teeth work like scissor blades, horses' back teeth grind like millstones.

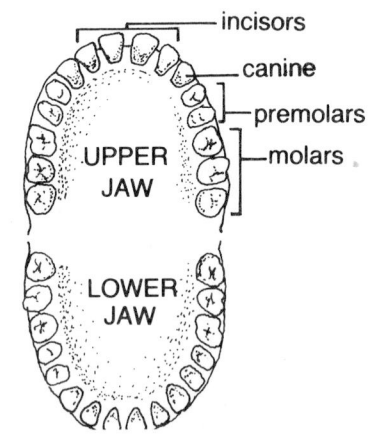

Plan of adult human teeth

ACTIVITY A.2
Copy the plan of a full adult set of teeth. Use a mirror to look at your own teeth or feel your teeth with your tongue or finger. On your plan, cross out any teeth which are missing from your mouth. Shade in any teeth which have fillings.

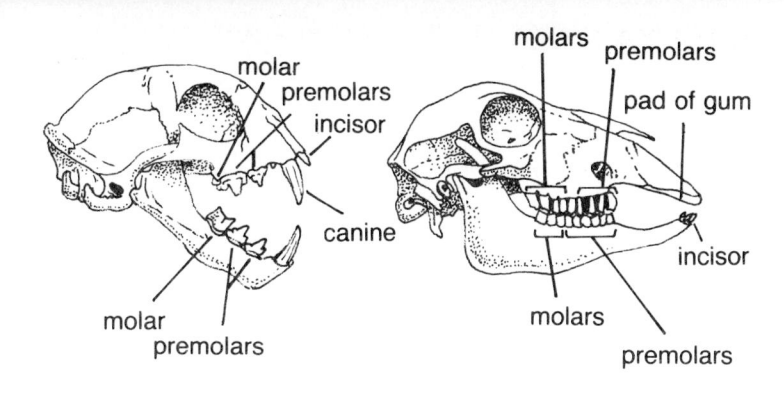

Cat's skull

Sheep's skull

Crocodile

Elephant

ACTIVITY A.3
Look at a skull of a herbivorous mammal, such as a sheep. Draw a side view of the skull. Label each kind of tooth. How does each kind play its special part in dealing with this animal's diet? Work out how the jaws move. How does the kind of jaw movement help to break up the food?
Look at a skull of a carnivorous mammal. Make drawings and notes as for the herbivore. What are the differences between the teeth in the two skulls?

Many animals use their teeth as weapons of defence, as well as for getting their food.

Some animals do not have teeth but they can bite. Birds use their beaks for biting. Many insects have biting jaws. These work sideways and cut off small pieces of food for the insect to swallow.

ACTIVITY A.4
Look at the biting mouthparts of a locust or a cockroach. You may be able to watch a live one feeding. You may be able to see the mouthparts of a dead insect mounted on a microscope slide. What food does this insect eat?

Sea urchins and starfish have a ring of teeth for biting and scraping at seaweed. Snails also scrape plants off stones. They have rows of tiny hooks on a tongue which works like a cheese grater or a carpenter's rasp.

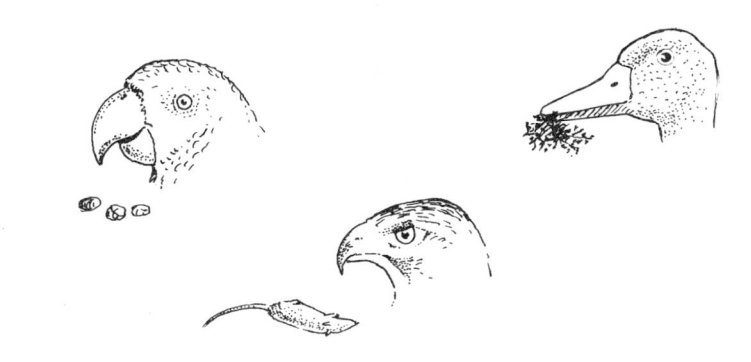

Birds' beaks
All birds have beaks. Each species of bird has a beak shaped and suited to the food that it eats. Birds use their beaks to pick up food and to break it. Some birds use their feet to help them catch or pick up food.

Owl

Humming bird

ACTIVITY A.5
Draw a picture of the head and beak of one species of bird. Write down what it eats. Explain how its beak is suited to deal with its food. Then choose a bird with different feeding habits. Make drawings and notes about its beak.

Sucking mouths
Some animals feed only on liquid food. A humming bird hovers near a flower. Its long, pointed beak is like a tube through which it sucks NECTAR. Some insects also have mouthparts like drinking straws, for sucking up liquid food.

ACTIVITY A.6
Look at the mouthparts of a mosquito, a housefly and a butterfly. You may have microscope slides of these, or you can use pictures.
1 What liquid food does a mosquito feed on? How is its mouth suited to getting at this food supply?
2 What liquid food does a butterfly feed on? How does it keep its long mouth tube out of the way when it is not feeding?
3 What does a housefly pour *down* its mouth tube? Why does it do this? What does it suck *up* its mouth tube? Why is it dangerous for us to eat food that a fly has been on?

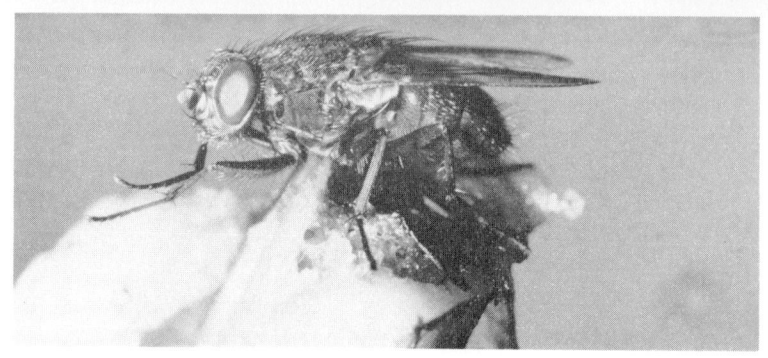

Housefly on meat

mosquito — mouth like a hollow needle for sucking blood

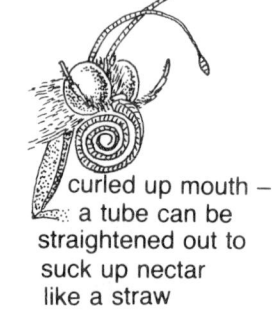

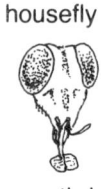

butterfly — curled up mouth — a tube can be straightened out to suck up nectar like a straw

housefly — mouth has a flat part for pressing on food and sucking juices

Mouthparts of insects

Catching food
Not all animals go searching for their food. Some animals stay still and wait for food to come to them. A frog may sit still until a fly comes near it. Then the frog shoots out its sticky tongue to catch it. The tongue can extend a long way because it is fixed at the front of the frog's mouth. Spiders spin webs to trap passing flies. Sedentary animals rely on currents of water to bring food to them. Sea anemones have stinging cells on their tentacles. These shoot out tiny harpoons to paralyse passing prey. The tentacles push this food into the animal's mouth.

Frog catching a fly

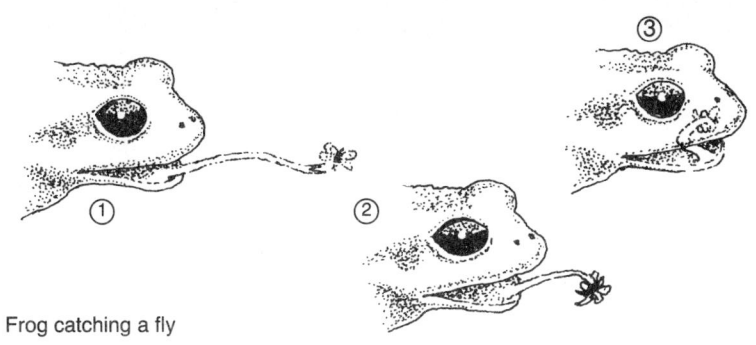

You may have seen barnacles at the seaside, forming a crust on rocks and piers. These animals are a strange kind of CRUSTACEAN. When they are covered by sea water, they push their legs out through a hole in the shell. The legs wave about, directing a current of water into the barnacle's mouth. The animal then filters the edible microscopic creatures out of the water, to eat them. Mussels are MOLLUSCS. They, too, live fixed to rocks in water. They also feed by filtering tiny creatures out of a current of water. Many aquatic animals obtain food by filter-feeding. The largest living animal, the blue whale, feeds by filtering tiny shrimps out of sea water.

Spider

Barnacle

ACTIVITY A.7
Choose one of each of the following:
1. an animal which traps its prey;
2. an animal which poisons its prey;
3. an animal which filters its food from water.

For each animal explain how it captures and eats its food. Explain how parts of its body are suited to its way of feeding. Make drawings of the animals feeding.

Parasites

Parasites are creatures which get their food from the body of a *living* host. Many children are 'hosts' to threadworms which live inside their intestines. These worms are a nuisance but fairly harmless and can easily be killed by proper medicine. Bigger roundworms and tapeworms living in intestines are more serious parasites of man and domestic animals. All these worms have no problem about feeding. They absorb the host's food which surrounds them. They do not need mouthparts for catching or breaking up food. Some have hooks or suckers so that they can fix themselves firmly inside their host.

Fleas, lice, ticks and mosquitoes are *external* parasites. They have to puncture their host's skin to feed. They have special mouths for doing this.

ACTIVITY A.8
Find out as much as you can about one species of animal which is a parasite. Make a poster about this parasite. Include information about the harm that it does, and how it can be controlled.

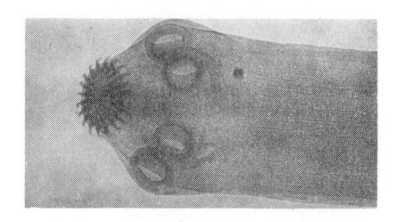

The head of a tapeworm showing its hooks and suckers

OPTION B

Animal Olympics

Imagine an Olympic Games of the animal kingdom. On land the animals are running a 200 metres sprint. The cheetah is obviously the champion. If these animals entered a 1 500 metres race, several kinds of antelope would overtake the cheetah. The 'big cats' cannot sustain their pace over a long distance. Man would be far behind the leaders in both events. Man fares even worse in the 100 metres free-style swim. The sailfish is champion of this

Speeds of animals

| | 20 | 40 | 60 | 80 | 100 | 120 | 140 | 160 km/hr |

AIR: bee, pelican, gull, owl, dragonfly, crow, golden eagle, wild turkey, Canada goose, peregrine falcon, swift

LAND: elephant, man, giraffe, horse, antelope, cheetah, tortoise, snail, earthworm, frog, cat, hare

WATER: man, polar bear, leather-back turtle, trout, gentoo penguin, sailfish, flying fish, eel, fin-back whale, tuna

event. Man is not even qualified to enter the flying race. The winner of this event, the spine-tailed swift, is the fastest of all animals, travelling at over 150 km/hr.

ACTIVITY B.1
Find out the answers to these questions:
1. To which *group* of animals do most of the leaders *on land* belong?
2. To which group do the ostrich and the emu belong? How do they differ from most members of this group?
3. What is the land speed record for animals, and which animal holds it?
4. Which groups of animals are the slowest on land?
5. What is the water speed record for animals, and which animal holds it?
6. Which swimming animals in the picture are *not* fish? In each case, to which group of animals do they belong?
7. What is the fastest animal in the world?
8. To what group do the fastest flyers belong?
9. Which other animal groups have members which can fly?
10. Why can man not fly unaided?

The champion athletes of the animal kingdom are fairly large animals. The thousands of species of tiny invertebrates are poor performers, if judged in straight competition. An animal high jump event would be won by a kangaroo jumping ten metres, with man jumping about two metres. A flea would jump less than 2 cm high. But if each animal were judged in comparison to its size, the records would be different. A man can jump about his own height, but a flea can jump 130 times its own height. If man could jump 130 times his own height he would easily clear the tallest building in Britain!

ACTIVITY B.2
Good jumpers belong to various animal groups. What do they all have in common which suits them to jumping?

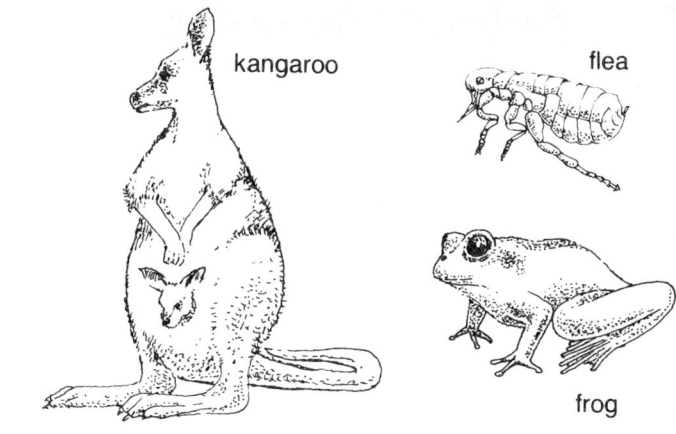

Some jumping animals

Hot-blooded champions

Birds and mammals are the absolute speed champions of the animal kingdom. They are HOT-BLOODED animals or ENDOTHERMS. An endotherm gets its body heat mainly from oxidizing its food. It can keep its temperature the same, whatever the weather. All other groups of living animals are COLD-BLOODED or ECTOTHERMS. An ectotherm gets its body heat mainly from its surroundings. Its temperature varies with the outside temperature. On a hot day a tortoise or a fly has a high body temperature. This makes everything happen quickly in the body and the animal moves fairly quickly. On a cold day these animals have a low body temperature which makes all their activities slow down.

The high, constant temperature of birds and mammals allows all their body activities to go on quickly and steadily. This high, constant METABOLIC RATE allows these animals to move rapidly and for long periods of time. They need a lot of food to provide heat energy to keep the body warm and to provide the energy for movement.

Lion, an endotherm

Lizard, an ectotherm

ACTIVITY B.3
Answer these questions.
1 Why is it easier to catch a fly on a cold day than on a hot day?
2 When and why do tortoises hibernate?
3 Is a cold-blooded animal's blood cold all the time? When would it be hot?
4 What is your normal body temperature?
5 How does your body keep its temperature down, on a hot day?
6 A lion is a mammal and an endotherm. A crocodile is a reptile and an ectotherm. They are about the same size. A lion has to eat far more food than a crocodile. Why is this?
7 Birds have feathers and mammals have hair or fur. Why are these coats needed by hot-blooded animals?

Man, the mechanical champion

If mechanical aids were allowed in the animals' Olympic Games, man would win every event. Man can go faster and further than any other animal, by using his machines.

Most animals are restricted to one or two ways of moving. Most are restricted to living in one place. Each animal's body is suited to its habitat and its way of life. A camel is suited in many ways to living in the desert. It has long legs and broad cushioned hooves for striding across sand.

Man has overcome the restrictions of his body. Man walks on two legs. This leaves his hands free to use tools and make aids to living and moving. But man's highly-developed brain is needed to think of helpful inventions. It co-ordinates his hands in making them. Man's brain and power of speech has also co-ordinated society so that men can communicate and work together. Some of man's inventions are copied from animal structures. Others are unlike anything in the animal world.

Concorde — with the people who built it

ACTIVITY B.4
Collect pictures of inventions which enable man to live and travel on land, sea and air, under the ground, under the sea and in space. Make a poster using your pictures.

ACTIVITY B.5
You have heard of Batman and Spiderman. Write a story about 'Moleman', 'Polar-bear-man', or 'Kangaroo-man'. First you must find out as much as you can about the movement and way of life of your animal. Then imagine how a man's body could grow to be like the animal's. Or think of an invention that an ordinary man could use to be like the animal.

OPTION C

Investigating your breathing

When you breathe you are pumping air in and out of your lungs. Breathing provides the oxygen needed for all your cells to respire. It also removes waste carbon dioxide from your body.

ACTIVITY C.1
Feel your breathing movements. Put your hands on the sides of your chest. Breathe in. This means draw air *in* through your nose. What is your chest doing? Breathe out. This means push air *out*. What is your chest doing? Does any other part of your body move as you breathe in and out? Feel your breathing movements when you breathe normally and when you breathe deeply. What differences can you feel?

The movement of your chest (or THORAX) pumps air in and out of your lungs. Between your ribs are INTERCOSTAL MUSCLES. At the base of your thorax is a sheet of muscle called your DIAPHRAGM.

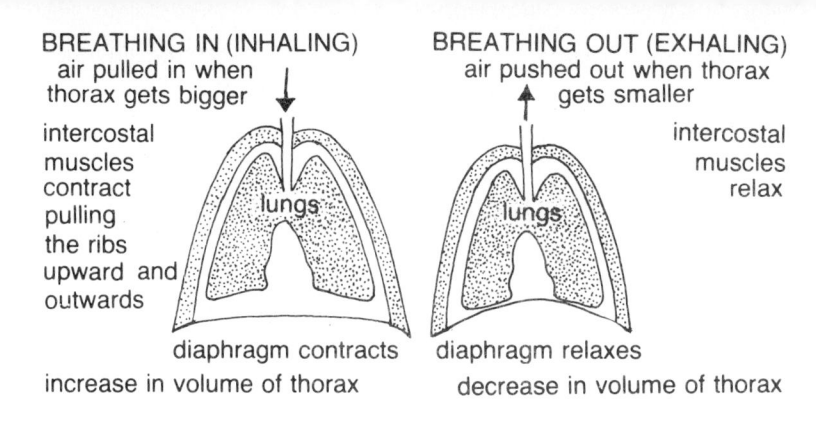

BREATHING IN (INHALING)
air pulled in when thorax gets bigger

intercostal muscles contract pulling the ribs upward and outwards

diaphragm contracts
increase in volume of thorax

BREATHING OUT (EXHALING)
air pushed out when thorax gets smaller

intercostal muscles relax

diaphragm relaxes
decrease in volume of thorax

ACTIVITY C.2
Use a model thorax made from a bell jar. Make the model 'breathe in' and 'breathe out'. How is this model like your breathing system? How is it different?

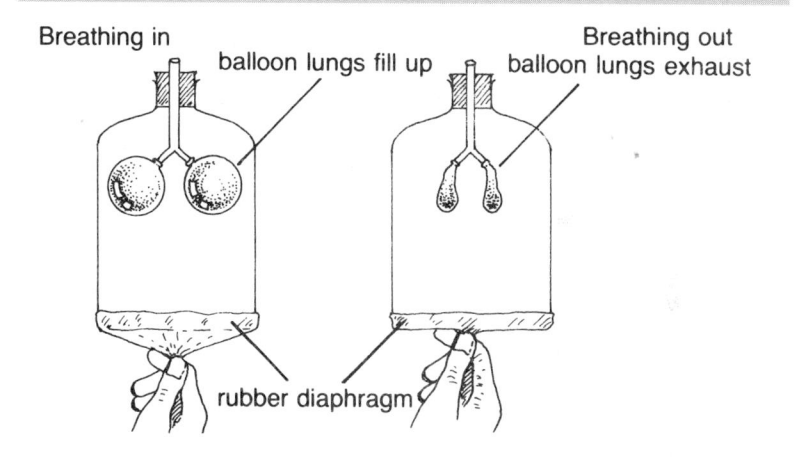

Breathing in — balloon lungs fill up
Breathing out — balloon lungs exhaust
rubber diaphragm

ACTIVITY C.3
Manipulate a jointed skeleton, or watch a film of breathing movements, to learn more about the position and action of the breathing muscles in man.

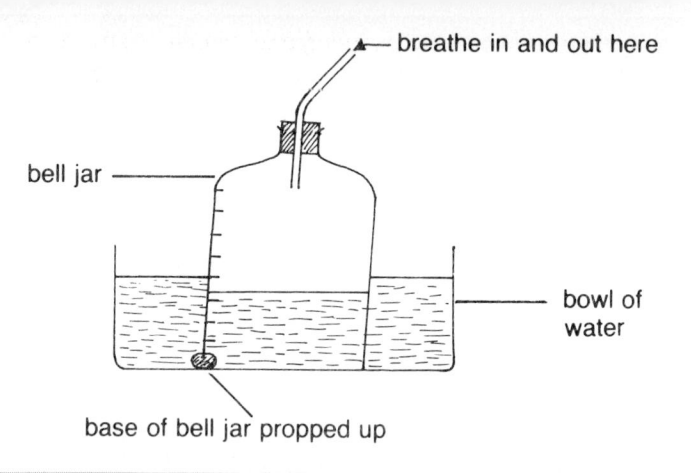

ACTIVITY C.4

Find how much air you breathe in and how much you breathe out. Work with a partner. One person breathes in and out through the tube of the apparatus shown in the diagram. His partner watches how far the water moves in the apparatus. The rise and fall of the water against the graduations measures the volume of air you breathe in and out. Measure the volume inhaled and exhaled in normal breathing, then repeat for deep breathing.

Changes in the air you breathe

The air you inhale is ordinary air. It is a mixture of gases, about 80% nitrogen and about 20% oxygen. There is a small amount of carbon dioxide, about 0.04%, and a little water vapour. The following Activities show how the air is changed by breathing.

ACTIVITY C.5

Collect a gas jar of exhaled air by displacement of water, as shown in the diagram. Have another gas jar of ordinary air. Fix short pieces of candle onto two deflagrating spoons. Light each candle and put one into each gas jar, quickly. Make sure the lid covers each jar quickly. For how long does each candle burn?

Can exhaled air support burning as well as inhaled air can? What differences might there be between inhaled and exhaled air, to explain your results?

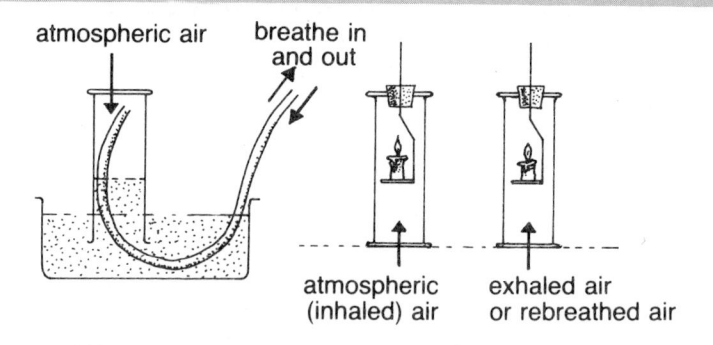

ACTIVITY C.6

Carbon dioxide can be detected by lime water. Lime water goes chalky white when enough carbon dioxide bubbles through it. The tiny concentration of carbon dioxide in ordinary air is not enough to change lime water in a short time.

You must use freshly prepared lime water. Breathe in and out gently through the mouthpiece of the apparatus shown in the diagram. Continue until you see a change in the lime water. What do your results show about the difference between inhaled and exhaled air?

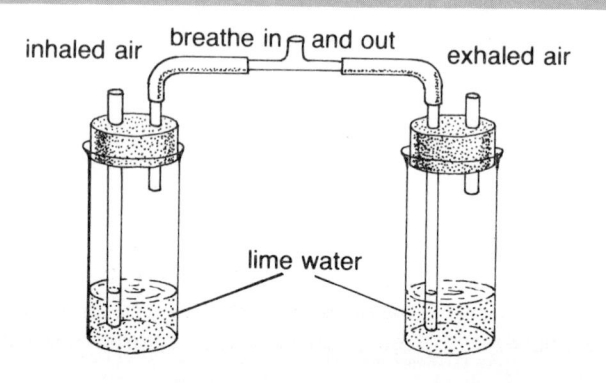

ACTIVITY C.7
Hold a mirror in the ordinary air that you inhale. Now exhale onto the mirror. What happens to the mirror surface? What does this tell you about a difference between inhaled and exhaled air?
To check whether an accident victim is breathing, you could hold a mirror near his mouth and nose. What would you expect to see if he were breathing?

Breathing and respiration

To the scientist, RESPIRATION means the release of energy in the cells of all parts of the body. BREATHING is needed to get oxygen into the body for respiration. So the breathing system is sometimes called the respiratory system. Artificially pumping air in and out of an unconscious accident victim is called artificial respiration.
If you are very energetic, your muscle cells need extra oxygen to release extra energy. Therefore you have to take in more oxygen and circulate it faster round your body.

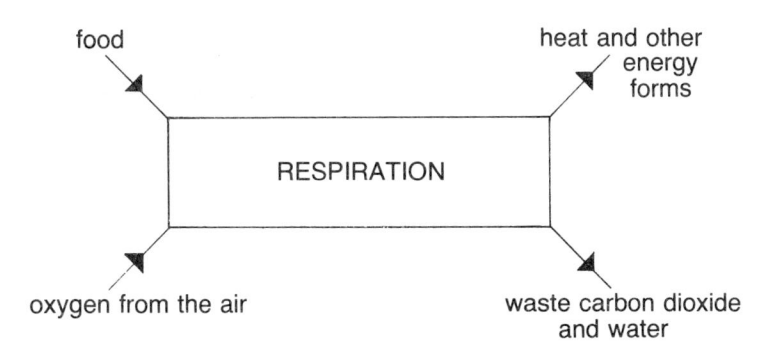

```
food                    heat and other
                        energy forms
        ┌─────────────┐
        │ RESPIRATION │
        └─────────────┘
oxygen from the air    waste carbon dioxide
                       and water
```

ACTIVITY C.8
For this experiment you have to take some strenuous exercise. Ask your teacher's advice about what you should do, e.g. go up and down stairs several times. Do the various activities in the table opposite. For each activity, your partner must count how many times you breathe in during one minute. He must count your pulse (heart beats) for one minute. He must notice how deeply you are breathing. Record all this in your table.

	Breathing rate	Depth of breathing	Pulse rate
Sitting at desk, breathing normally			
Lying down for five minutes			
Standing up			
Immediately after exercise			
Ten minutes after finishing exercise			

Which activity requires the most energy? How did your body adjust to help to supply the energy? Why did you take a while to recover from exercise? Why were there differences between the results for sitting, lying and standing?

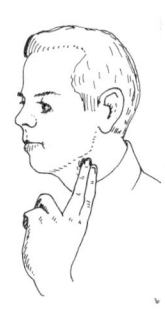

Taking a pulse

OPTION D — Measuring growth

Animals grow slowly. To study growth we must take careful measurements over a long time. Most of the Activities in this Option have to be started straight away and continued over several weeks or months. The feature which we choose to measure is a CRITERION of growth. The measurements or information you collect are DATA. You must RECORD your data carefully in tables and/or graphs. By studying your records you will be able to GENERALIZE about your growth criterion.

ACTIVITY D.1
Make a scratch just above the base of your thumbnail. Use a clean blade or needle supplied by your teacher. Measure in millimetres the distance of the scratch from the nail base. On the same day every week, for about ten weeks, measure this distance. How fast does your nail grow (in mm per week)?

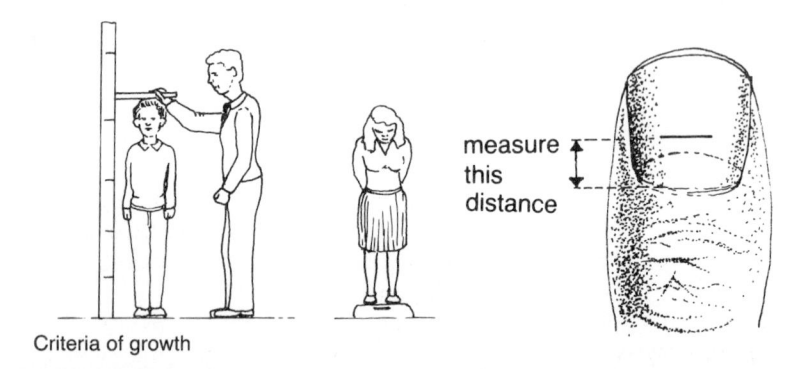

Criteria of growth

ACTIVITY D.2
Use height and weight as criteria for judging your overall growth. Measure your height and weight on the same day every week. Wear the same amount of clothing each time. Continue for a whole school year if possible. Plot graphs of your height and your weight against time. Compare your results with other people's.

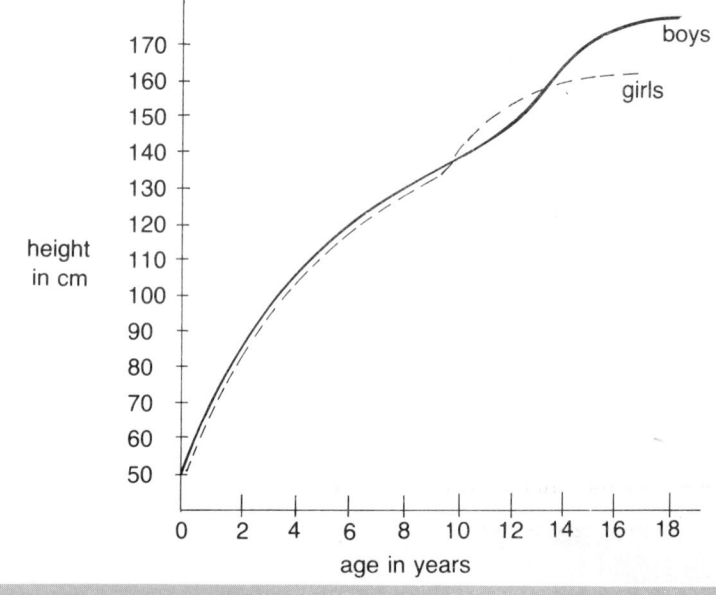

ACTIVITY D.3
Look at the graph of boys' and girls' heights. On average, are boys or girls taller at birth; at 13; at 16? When do most girls complete their growth in height? Can you tell from the graph when most boys complete their growth in height? Do people grow at a regular rate, or in spurts?

You may find that you are much shorter or much taller than the average for your age. This is nothing to worry about. There is a wide variation of normal heights and weights at any age. In the teens there is a particularly wide range in each year of age. Some people have their ADOLESCENT GROWTH SPURT or stage of rapid growth at the beginning of their teens so they are tall as teenagers. Others are small through their middle teens, then have their growth spurt at seventeen or older. They may even overtake the early developers.

ACTIVITY D.4
Make a survey of the height of pupils throughout your school. Work out the average height of boys and the average height of girls in each year group. If your school has a limited age range, or only one sex, perhaps you can collect height data about other ages and the opposite sex from neighbouring schools. Plot graphs of average height against age group, with girls and boys separately, as in Activity D.3. Do your results correspond with the graph shown here?

ACTIVITY D.5
Measure the growth of a small mammal, such as a gerbil or a rat. Choose one or more criteria of growth, e.g. weight, tail length. Start as soon as you can safely handle the baby animals. Take the average measurement from several individuals if possible. Continue until the animals are full-grown.

A gerbil

ACTIVITY D.6
Look at the table of data about gerbil growth, collected by a pupil during one school term. Calculate the average gain in weight for each week. Did the gerbils grow at a steady rate? When did they stop growing? Did both stop growing at the same week? Why did the average weight fluctuate during the last three weeks of the experiment? Compare these results with ones you obtained in Activity D.5.

Date	Week number	Average weight (g)	Date	Week number	Average weight (g)
9 Sept	0	10.3	4 Nov	8	42.0
16 Sept	1	15.4	11 Nov	9	44.2
23 Sept	2	18.6	18 Nov	10	50.0
30 Sept	–school	holiday–	25 Nov	11	55.0
7 Oct	4	27.9	2 Dec	12	61.0
14 Oct	5	32.6	9 Dec	13	59.0
21 Oct	6	34.0	15 Dec	14	60.5
28 Oct	–school	holiday–			

Gerbil growth. Two gerbils were weighed weekly (except on holidays)

ACTIVITY D.7
Study the growth of an animal which undergoes metamorphosis. Find some frog spawn or tadpoles, butterfly eggs or caterpillars. Keep your animals in suitable surroundings, supplied with suitable food. Watch their development. What could you *measure* as criteria of growth of your animals? Would the same criteria be suitable for all kinds of animals, and at all stages of their development?

CHAPTER 5

Man's place in the animal kingdom

ACTIVITY 5.1
Look again at your classified list or collection from Activity 1.3. Do you want to alter your classification now that you have learnt more about animals? Discuss it with your friends.

Scientists classify animals according to overall similarities. This is based on their ANATOMY (body structure) rather than their way of life or their HABITAT (where they live). This natural system of classification shows us the pattern of evolution of animals.

You have a backbone and other bones making up your skeleton. Animals with backbones are called VERTEBRATES. The vertebrates make up one big group, separate from the animals without backbones, the INVERTEBRATES.

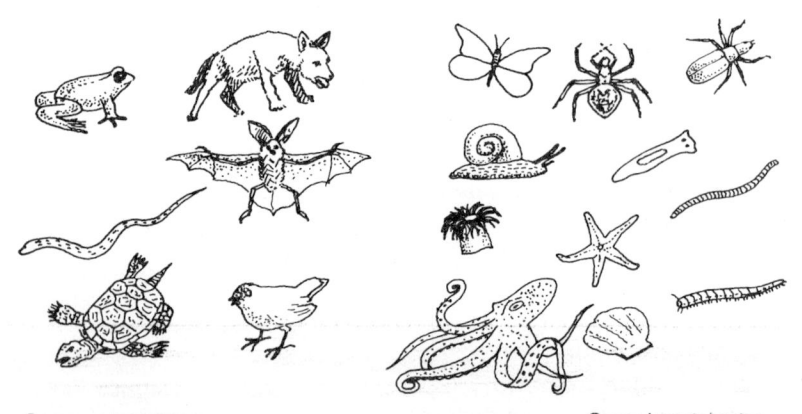

Some vertebrates   Some invertebrates

Invertebrates
The animals without backbones are very varied. Scientists divide them into groups, each called a PHYLUM. All the animals in a phylum are alike in some important ways. The diagram on page 31 shows you some features of the members of each phylum.

ACTIVITY 5.2
Classify your collection again. For each animal, first decide whether it is a vertebrate or an invertebrate. Then decide which phylum each invertebrate belongs to. Add examples so that you have listed some members of every phylum.

Invertebrates with jointed legs
Every phylum is subdivided into smaller groups. The phylum ARTHROPODS contains about ¾ million species. These are classified into groups according to their number of legs.

ARTHROPODS
(several pairs of jointed legs)

INSECTS
(3 pairs of legs)

ARACHNIDS
(4 pairs of legs)

CRUSTACEANS
(between 4 and 20 pairs of legs)

MYRIAPODS
(More than 20 pairs of legs)

ACTIVITY 5.3
Classify the arthropods in the diagram on page 31 into their correct groups.
Make a large wall chart called 'Classification of the Arthropods'. Use reference books to find out further subgroups of insects, crustaceans, etc. Cut pictures out of old magazines, or draw animals, to illustrate each group.

THE ANIMAL KINGDOM

INVERTEBRATES

PHYLUM PROTOZOANS
Animals whose bodies are made of only one cell
e.g. amoeba

PHYLUM COELENTERATES
Animals whose bodies are jelly-like bags with only one opening. They have stinging cells on tentacles
e.g. hydra

sea anemone

jellyfish

PHYLUM FLATWORMS
Flat, leaf-shaped or tape-shaped animals. Many are parasites
e.g. planarian
tapeworm

PHYLUM SEGMENTED WORMS
Animals with tubular bodies made of ring-shaped segments
e.g. earthworm

leech

PHYLUM MOLLUSCS
Animals with one or two shells protecting a muscular foot
e.g. snail

mussel

octopus

PHYLUM ECHINODERMS
Animals whose bodies are arranged on a plan of five equal parts radiating from a mouth
e.g. starfish

sea urchin

VERTEBRATES

PHYLUM ARTHROPODS
Animals with a hard skin forming an external skeleton. They have jointed legs
e.g. housefly

butterfly

crab

woodlouse

spider

scorpion

centipede

shrimp

mite

Back to backbones

Animals with backbones are classified into the following groups: Fishes, Amphibians, Reptiles, Birds and Mammals. These groups have many distinguishing features. The skin of a vertebrate gives a good clue to its group. Fishes have wet, scaly skin. Amphibians have smooth, moist skin. Reptiles have a dry, scaly, leathery surface. Birds' skin grows feathers and mammals' skin grows hair or fur.

ACTIVITY 5.4
Classify the vertebrates on your original list into their groups. Make a large illustrated chart called 'Classification of Vertebrates'. Find out how the main groups are subdivided.

Mammals
Mammals are warm-blooded, hairy vertebrates. Their young grow in the mother's womb and are born as babies, not eggs. The babies feed on milk from their mothers' mammary glands.

ACTIVITY 5.5
Make a list of as many mammals as you can. Subdivide mammals into groups (ORDERS). What animals would you put into the same order as man?

Monkeys, apes and man

Scientists put lemurs, monkeys, apes and man into an order called PRIMATES. These animals all have eyes on the front of their faces, well-developed brains and hands which can grasp objects. They care for their young for a long time after birth, often in family groups.

Biologically, man is an animal. You are classified as a member of the species *Homo sapiens*, in the primates order of mammals. Man's brain, eyesight, grasping hands and parental care are more developed than those of any other primate.

ACTIVITY 5.6
What things can man do that no other animal can do? Do these activities make man a special kind of animal, or make him completely different from animals? Write your ideas about this, or have a class discussion.

Most animals are biologically adapted to living in one particular environment. Man is more successful than other animals because he can change his environment. For example he can build houses which have central heating or air-conditioning. One individual man could not alter his environment a great deal. People live and work together to exploit their environment to everyone's advantage. Man is a SOCIAL ANIMAL. Our society is much more complicated than that of bees, ants, chimpanzees or other social animals.

All social animals have DIVISION OF LABOUR. This means that different individuals are specialized to do different jobs to contribute to the life of the community. In a bee colony, queen, drones and workers are biologically different to fulfil different functions. In human society, people are biologically similar but they are trained

Human activities: using tools, painting, speaking, praying and writing

in different skills to provide division of labour. In all species of social animal, individuals have to communicate with each other. Humans can communicate better than any other animal species, because they can speak, write and use technology.